TIGER in action

By Bruce Culver

Color by Don Greer
Illustrated by Perry Manley

Armor Number 27
squadron/signal publications

Tiger ausf Es of the 13th *Kompanie, Leibstandarte Adolf Hitler* move across the Russian countryside during the battle of the Kursk salient in July of 1943.

If you have any photographs of the aircraft, armor, soldiers or ships of any nation, particularly wartime snapshots, why not share them with us and help make Squadron/Signal's books all the more interesting and complete in the future. Any photograph sent to us will be copied and the original returned. The donor will be fully credited for any photos used. Please send them to:

Squadron/Signal Publications, Inc.
1115 Crowley Drive.
Carrollton, TX 75011-5010.

Acknowledgements

Jon Randolph   John W. Kennon
Paul Hocking   Armin L. Sohns
Matt Detrick   National Archives and Records Service

And, especially, Miles Krogfus, who gave of his time and knowledge to make the information on camouflage and markings the most accurate to date, and who also assisted in identifying many of the units and campaigns covered in the photographs. Without his help, this book would have been much less complete.

Dedication

To all who fought in the Tiger and to all who fought against it.

Photo Credits

U.S. Army   Squadron/Signal Archives
Bill Hess   Norman W. Kuhns
Richard F. Grant   Bundesarchive Koblenz

A Tiger I of the 101st SS sPzAbt advances across open country during the spring of 1944. The reputation of the Tiger was legendary and, for over two years, there was no other tank that could match its firepower, protection, and performance. This vehicle is camouflaged in Dark Yellow, Red Brown, and Olive Green, with the turret numbers in light Blue with Yellow outlines.

Introduction

Of the various weapons used during the Second World War, a number became famous and a few became household words, known to even those with no special interest in the weapons of war. Spitfire, Zero, Mustang, and a few others, are quickly recognized by a great number of people, even if the exact aircraft type is forgotten. In the West, there are probably only two tanks that ever reached this level of public familiarity; the American Sherman and the German Tiger.

The Tiger tank became a legend, partly through the efforts of the German propaganda machine, but, more importantly, because for nearly three years it was the most feared and respected of all German tanks. For over two years, no Allied tank could match its capabilities. Its combination of armor protection, firepower, and maneuverability made the Tiger the "Queen of Battle" and, even at the end of the war, there were few Allied tanks that could challenge a Tiger on even terms.

Work on the tank that would eventually emerge as the Tiger began during 1937, when the firm of Henschel und Sohn in Kassel, Germany, was directed to design a 30-33 ton tank as a replacement for the PzKpfw IV (which was just entering service). After one test vehicle was built, work was suspended on the prototype DW I (*Durchbruchswagen* -breakthrough vehicle) to allow Henschel to shift its assets to work on the VK.6501, a 65-ton heavy tank which was a progressive development of the pre-war *Neubaufahrzeuge* heavy tanks. In the event after construction of two prototypes, this project was cancelled and work was resumed on the DW I.

By 1940, Henschel had improved the DW I design and had given it the company designation, DW 2. The DW 2 weighed 32 tons, had a crew of five men, and carried an armament of one 75MM L/24 howitzer and two MG-34 machine guns. Trials with the DW 2 continued until 1941 when Henschel received an order for a new design of more advanced construction. This tank was designated the VK.3001(H) (*Vollkettenkraftfahrzeug* - full track experimental vehicle, 30 tons, first design in this class, built by Henschel). Daimler-Benz, Porsche, and MAN also received developmental contracts for prototype vehicles of this new class of tank.

The Henschel VK.3001(H) was the direct ancestor of the Tiger tank series, and was developed directly from the earlier DW 2 experimental vehicle. The superstructure followed the German practice in using vertical plates of welded armor bolted to the chassis and lower hull. The suspension consisted of seven axles, mounting interleaved road wheels which were sprung on torsion bars, with three return rollers on each side.

It had been intended to arm this tank with the 75MM L/48 high-velocity tank gun; however, the appearance of the Soviet T-34 tank made this weapon obsolete. The 75MM gun could not pierce the armor of the T-34, so it was decided to up-gun the VK.3001(H). Two of the VK.3001(H) chassis were converted into self-propelled gun carriages for the 128MM K L/61 anti-tank gun and were sent to Russia where they were used successfully against enemy armor. Because it was the direct ancestor of the Tiger, the VK.3001(H) was popularly known at Henschel as the *kleiner Tiger* (Tiger cub).

Concurrently with the development of the VK.3001(H), Hitler requested German tank manufacturers to design a still heavier tank which could dominate any battlefield. The specification called for a tank carrying a weapon capable of penetrating enemy tank armor at ranges out to 1,500 meters, sufficient armor protection to stop any known anti-tank weapon at the same distance, and a speed of at least 40 kph (25 mph). The vehicle's main weapon was to be the Geraet 725, a tapered bore gun firing tungsten core ammunition. The VK.3601(H) had excellent armor protection with 100MM armor in all front plates, and 60-80MM on the sides and rear. At the time it was ordered, this vehicle would have been extremely difficult to destroy with any of the current Allied anti-tank weapons.

The chassis was developed from the VK.3001(H), but used eight axles with larger road wheels, and the return rollers were deleted.

In the event, development of the Geraet 725 gun was abandoned due to a shortage of tungsten. Without adequate supplies of tungsten, there would be shortages of ammunition for combat units. The only alternative weapon in the same class was the excellent 88MM KwK36 cannon, which had been developed from the Flak 18 dual-purpose anti-aircraft gun. Henschel agreed to develop a new tank to mount the KwK36, under the company designation VK.4501(H).

The VK.4501(H) was adapted directly from the VK.3601(H) and was very similar in design and appearance. The major change between the two was in the hull which was extended outward above the tracks. This extension was necessary to allow for the installation of the KwK36, which required a turret ring diameter much larger than that proposed for the 75MM gun originally intended for the VK.3601(H). The superstructure side armor was increased to 80MM thickness. A new turret was developed by Rheinmetall for both the VK.4501(H) and Porsche's entry the VK.4501(P). Rheinmetall also built a wood mockup of an alternate turret for the Henschel design capable of mounting the 75MM KwK L/70 cannon; however, this was abandoned, and the L/70 gun was eventually adapted for use use on the Panther medium tank.

One difficult stipulation of the VK.4501 specification was that the first prototype vehicles had to be ready for Adolf Hitler to inspect them on 20 April 1942, Hitler's birthday. Since the order was dated 26 May 1941, this left less than a year to design and construct prototypes capable of being fully evaluated. This placed the engineering staff and construction workers under enormous pressure. By working day and night, Henschel's work crews managed to finish the VK.4501(H) prototype by 17 April 1942, only forty minutes before it had to be loaded onto a train for transport to Hitler's headquarters at

The unsuccessful Porsche VK.4501(P) prototype rests on a railway flatcar during the vehicle's trials. Compared to the Henschel VK.4501(H) prototype, the Porsche vehicle was unmaneuverable and the electric drive system experienced a number of problems. The vehicle was painted overall Dark Gray. (Bill Hesz)

Rastenburg. There was not even time to test drive the vehicle which had to be taken to the rail yard on a trailer for loading.

On 19 April 1942, both the Henschel and Porsche VK.4501 prototypes arrived at the unloading point some 11 km (7 miles) from Rastenburg. They were both off-loaded from their flatcars by crane, and the Porsche vehicle quickly bellied into the soft ground. Kurt Arnoldt, Henschel's chief engineer, offered Dr. Porsche the use of the Henschel prototype to tow the the stuck VK.4501(P) prototype, an offer that Porsche brusquely refused.

In fact, neither vehicle was truly ready to drive. The Henschel entry had never even moved under its own power! The 11 km drive to the *Fuehrerhauptquartier* at Rastenburg must have seemed an eternity to the two competitors. Neither vehicle could move more than a few hundred meters without breaking down, and both tanks were overhauled several times along the route. At one point, Kurt Arnoldt had to drive pins into the troublesome final drive mechanism of the Henschel prototype to keep the tank running.

The next day, 20 April, both prototypes were made ready for Hitler's inspection. When word was relayed that Hitler would be delayed until two o'clock that afternoon, Arnoldt ordered the VK.4501(H) final drive torn down for repairs. As luck would have it, Hitler showed up with his entourage at 11 AM, causing great confusion in Henschel's area as Arnoldt and his workers attempted to quickly reassemble the final drive. Both tanks were to make a high speed drive of several hundred meters to demonstrate their speed capability. The Porsche vehicle made a successful run of over 1,000 meters at a speed of 50 kph (30 mph). Arnoldt drove the Henschel entry, which went about 850 meters at 45 kph (25 mph). On the return trip, the engine heat built up so much Arnoldt feared that the tank might catch fire. He stopped short of the reviewing group and managed to get away with this, while hoping the vehicle would cool down before anyone had to drive it again.

The somewhat superior speed performance of the Porsche prototype had led some members of the tank industry to belittle Henschel's entry. Arnoldt approached Albert Speer, however, suggesting a test of maneuverability. He knew that the Porsche vehicle was very hard to turn, and also that the regenerative steering final drive gearbox in the Henschel tank gave the best maneuverability of any large tank in existence. The resulting test was a success for Henschel and, because of this and many other design problems with the Porsche prototype's electric drive transmission and chassis, the Henschel prototype was chosen for series production. The Porsche chassis was later adapted for use as a self propelled anti-tank gun, known as the *Ferdinand* (later *Elefant*).

The Henschel VK.4501(H) was ordered into production under the service designation PzKpfw VI Tiger ausf E, SdKfz 181; however, it was generally called Tiger by most of the troops using it, and by the Allies opposing it. With the advent of the later Tiger II, some sources then referred to the original model as the Tiger I, though this name was never officially given to the vehicle.

TIGER Prototypes

**Porsche
VK.4501(P)**

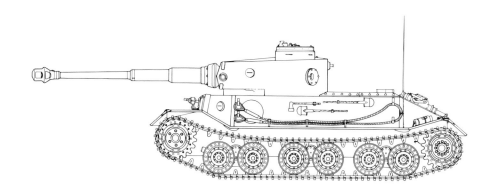

**Henschel
VK.4501(H)**

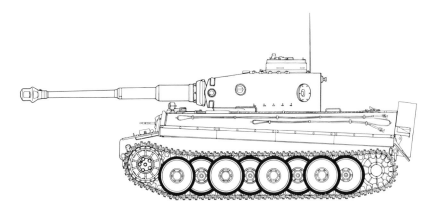

Development

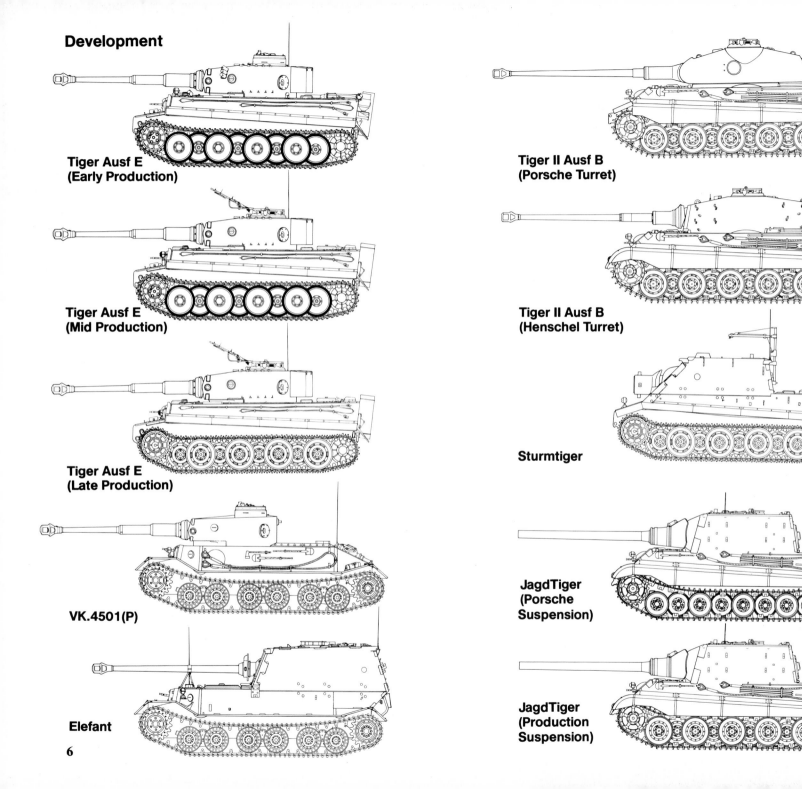

Tiger Ausf E (Early Production)

Tiger Ausf E (Mid Production)

Tiger Ausf E (Late Production)

VK.4501(P)

Elefant

Tiger II Ausf B (Porsche Turret)

Tiger II Ausf B (Henschel Turret)

Sturmtiger

JagdTiger (Porsche Suspension)

JagdTiger (Production Suspension)

Tiger Production

Henschel und Sohn, of Kassel, Germany, was a well-known manufacturer of heavy industrial and railroad equipment, especially railroad locomotives and large dock cranes. Because of the size and weight of the Tiger, Henschel was considered to be the ideal manufacturer, having all the facilities needed to produce such a heavy vehicle. Henschel also had a fine engineering staff, led by Kurt Arnoldt, and a complete vehicle test facility. The final assembly hall at Henschel's plant dwarfed the tanks being produced there and the final assembly line was capable of producing several tanks each day.

Although much of the installed equipment on the Tiger was subcontracted, Henschel manufactured most of the major components in their plant. Hulls, turrets, and other contract items and assemblies were brought into the assembly building where final machining operations and detail assembly were done. Henschel's facilities allowed the firm to machine the turret rings and other critical areas of the hull within the plant without outside assistance.

Crews for the Tiger were hand picked, both from operational units and from the top trainees at the *Panzerschulen* (tank training schools). These crews were sent to Henschel to learn about the detail workings of the Tiger, and to watch their vehicles undergoing final assembly. It was an excellent idea, since the Tiger required a great deal of preventive maintenance, and a well-trained crew was necessary to keep the vehicle operational. One of the training aids was the *Tigerfibel*, a humorous illustrated booklet that gave hints and tips on keeping the Tiger operating properly. It was distributed to all new Tiger crews and was a popular part of the training process.

Lowering the turret onto the hull was done near the end of the assembly process, after the powertrain and major internal components had been installed. The lower suspension and hull of this Tiger are already painted in Dark Yellow. The mudguard on the port side has been folded up, as was done to prepare the tank for rail transport.

Henschel produced heavy machinery such as railroad locomotives and large cranes, and was well suited to build the Tiger. The large assembly hall was very well equipped for the machining operations needed to produce tank hulls. After the hull was welded together, the suspension and turret openings were machined to exacting tolerances, after which final assembly could begin.

During April of 1944 Tiger production reached its peak when 105 vehicles rolled off the assembly line. Total production of all Tiger ausf E vehicles was 1,354. These were delivered from July of 1942 until August of 1944. Shortages of some materials, changing war conditions, and the introduction of the Tiger II ausf B during January of 1944 were some of the factors that limited Tiger production to the original contract placed during early 1942. Although the Tiger was very maneuverable for its weight and size, it was not as fast or agile as lighter tanks, such as the Panther. Additionally, it was difficult and expensive to manufacture, and the use of pre-war type vertical armor limited the amount of protection the Tiger could carry and remain near its weight limit. If fact, production Tigers exceeded the original prototype's forty-five ton weight limit by as much as eleven tons.

While the Germans stayed close to their original production schedule for the Tiger, it is interesting to note that the American T26E3 (M26) Pershing tank was built at a rate of 1,350 tanks over a six month period. In the end, it was this difference in production philosophy and faster American production that made the difference between defeat and victory.

7

A new production Tiger leaves the final assembly building at the Henschel plant. This is a mid-production vehicle; however, it retains many of the main features of early Tigers. The small angled tubes mounted on the hull top are "S" anti-personnel mine dischargers.

The bolts on the hull sides are used to attach the side skirts. The narrow transport tracks were required to avoid excess overhang during rail shipment of Tigers to the front. Many tunnels and bridges were too narrow to allow a fully tracked Tiger to clear, especially on curves.

Tiger ausf E

Production Tiger Ausf E tanks differed very little from the prototype with the exception of having a large two-compartment external stowage bin mounted on the rear of the turret. Each compartment had its own hatch and the bin was notched in the center to clear the rear turret lifting hook.

Production of the Tiger ausf E followed standard German construction guidelines. The hull and superstructure were of welded construction using vertical armor plates. The Tiger's armor featured interlocking plates that strengthened the joints. The hull and superstructure front armor were 100MM thick, as was the flat gun mantlet on the turret. The hull sides were 60MM thick and the superstructure sides and rear plate were 80MM thick. The 25MM thick hull roof plate was recessed and welded to the superstructure sides.

The turret was formed from a single piece of 82MM thick armor bent to the shape of a horse shoe, with the two small front plates that completed the turret being 100MM thick. The turret roof plate was 25MM thick and was recessed and welded to the sides of the turret. The roof was also bent, forward of the center line, to follow the tapered contours of the turret sides.

The interior of the Tiger was divided into four major compartments: driver, bow gunner/radio operator, turret fighting compartment, and engine compartment. The two front compartments were actually one hull space divided by the massive transmission and final drive unit. The driver sat on the port side, and steered by means of a large control wheel which operated the hydraulic power steering system. As a back-up, two manual emergency steering levers were provided should the power steering system fail. These levers operated the manual steering disc brakes, which were also used as parking brakes. The driver's armored visor was protected by vertically moving blocks which were operated by a handwheel on the front plate.

This Tiger is probably intended for duty in North Africa or Russia and is equipped with Feifel air cleaners mounted on both sides of the rear engine deck. The brackets on the turret side are for holding spare track links and the small cable on the hull side is used for changing the transportation and battle tracks.

The radio operator sat on the starboard side of the hull and also served as the bow machine gunner. The bow machine gun was a 7.92MM MG 34 housed in a standard *Kugelblende* gun mount installed in the front armor plate. The radio sets were mounted on a shelf to the left of the radio operator, over the transmission.

A substantial arched cross brace, which supported the weight of the forward turret, also served to separate the turret fighting compartment from the forward area. The massive turret sat on a large turret ring 1.85 meters (73 inches) in diameter, and the turret floor was suspended from the turret ring by three support tubular struts. The breech mechanism of the 88MM KwK36 gun extended almost to the rear of the turret, effectively dividing the turret into two sections. The gunner sat to the port side of the gun breech and was provided with a TZF9b binocular gun sight. Two small holes were drilled in the gun mantlet for the sight optics. Hydraulic power traverse was used for large movements of the turret, with fine adjustments being accomplished by use of a traversing handwheel. Elevation of the KwK36 main gun was done by use of a handwheel.

Aiming and firing the main gun took skill and concentration. The gunner had a dial-type indicator showing the clock position (1-12) of the main gun to aid in rapid aiming of the gun and for picking up new targets during the confusion of battle. The loader sat on the starboard side of the gun, and was responsible for loading the KwK36. Ammunition

for the gun was stored in bins beneath the turret basket, in the hull floor, and in the side hull superstructure sponsons.

The tank commander sat at the port rear side of the turret, below the cupola. The cupola on the early and mid production Tigers was a cylindrical drum shape, with a hinged top hatch and five armored glass vision slots around the circumference. A turret traverse indicator ring was fitted to the cupola, and gave the exact clock position of the main gun, duplicating the gunner's indicator. The cupola also had a mount for a rangefinder and a swinging arm mount for a standard scissors-type artillery spotting telescope.

The main armament of the PzKpfw VI Tiger ausf E was the 88MM KwK36 L/56 cannon. To allow for easier and faster elevation and depression of the gun, the weight of the gun was balanced by a counterbalance spring which was housed in a cylinder on the forward starboard side of the turret wall. The weapon was fired electrically from a switch on the gunner's traverse handwheel.

Stowage bins and racks held a total of ninety-two rounds of high explosive (HE) and armor piercing (AP) ammunition for the KwK36. Additional 4,800 rounds of 7.92MM ammunition was carried for secondary armament which consisted of two MG 34 machine guns. One gun was mounted in the front hull and the other was coaxially mounted in the main gun mantlet. For close in crew defense, a 9MM MP40 machine pistol was carried

New Tiger battalions were trained in France, where German armored units established a number of training areas. The crews learned how to best use the Tiger in unit sized maneuvers.

These tanks are overall Dark Yellow and carry no unit markings. During training, the guns were often covered to keep them clear of dirt and debris.

inside the hull. This weapon could be fired from pistol ports, one of which was located on the port rear side of the turret and the other on the starboard side.

Early in Tiger production, the starboard pistol port was replaced with a large round crew escape hatch which was hinged at the bottom and opened downward. Six NbK 39 90MM smoke dischargers were mounted on the sides of the turret, three on each side.

For defense against infantry attacks, a number of Tigers had individual anti-personnel "S" mine dischargers mounted around the superstructure roof at various points. The "S" mine consisted of a round bomb about five inches deep by four inches wide that contained some 360 steel balls. In operation, the discharger would fire the mine about three to five feet into the air, where it would explode, scattering the steel balls in all directions. The mounting of "S" mine dischargers was discontinued on later production Tigers with these weapons being replaced by the *Nahverteidigungswaffe* (close in defense weapon). The *Nahverteidigungswaffe* was an internally mounted and loaded anti-personnel mine thrower which could be traversed to counter any threat to the tank.

The Tiger's weight made it necessary to use interleaved road wheels carried on overlapping torsion bars to obtain the proper ground pressure. The right hand axles trailed the torsion bars; the left hand axles led the torsion bars. As a result, the road wheels on the Tiger were not offset as they were on most torsion bar suspension vehicles. To improve weight distribution, there were three road wheels on each axle, giving the Tiger a very stable and comfortable ride for such a large tank.

Two sets of tracks were issued with each vehicle: a wide battle track which was 725MM (28.5 inches) wide, and a narrow transport track which was used for rail shipment. The transport track measured 520MM (20.5 inches) in width. When the narrow track was fitted, the outer road wheel was removed from each axle. This was required to meet the tight clearances on many of Germany's railroad lines. In open country, whenever possible, Tigers were shipped with the battle tracks installed to save time in getting the vehicles ready for action.

Early production Tigers were fitted with the 650 hp Maybach HL 210 P45, a 21 liter displacement V-12 gasoline engine, mounted in the rear of the hull. From the end of 1943, production Tigers mounted the more powerful 700 hp Maybach HL 230 P45 engine. This engine had an enlarged displacement of 23 liters. Although the Tiger developed a reputation for being underpowered, the engine proved to be fairly reliable unless overstressed by attempts to tow other disabled tanks. The lack of suitable heavy tank recovery vehicles often led Tiger crews to tow disabled Tigers, sometimes resulting in the loss of both vehicles.

The final drive, mounted in the forward hull, had been developed by Henschel and was based on the pre-war British Merritt-Brown regenerative steering unit. Combined with the preselective eight-gear transmission, this new system made the Tiger easy to drive and far more maneuverable than any other heavy tank of the period. A major design tradeoff in the Tiger was the fact that the turret had to be removed to do any major repairs to the transmission or final drive, or to replace either of these units. The design of the Tiger's hydraulic system, however, made removal of the turret simple. To remove the turret, a simple dog clutch was disengaged, which disconnected the turret power takeoff from the transmission. The turret's hydraulic system itself was completely self-contained, being driven by the transmission power takeoff.

The first 495 Tigers off the production line were fitted with a snorkel system and seals to allow the tank to wade streams up to 4 meters (13 feet) deep. It is doubtful that this was ever used extensively, and later Tigers were instead fitted with wading equipment that allowed them to wade streams up to 1.3 meters (4 feet) in depth. Tigers intended for duty in tropical or dusty operating areas were fitted with the Feifel air cleaner system. These special air filters were mounted at the rear of the Tiger's hull and connected to the engine through metal ducts and flexible hoses mounted on the rear engine deck.

The antenna mast for the radios was mounted on the rear starboard side of the engine deck and usually mounted a rod type antenna.

This new production Tiger crosses a road during unit training exercises held in early 1943. It usually did not take long for the skirts and mudguards to begin to show signs of damage caused by running through brush and trees. Often, even heavier external parts were damaged by repeated heavy use and maintenance.

Turret Stowage Bin

Tiger Prototype

Production Tiger Ausf E

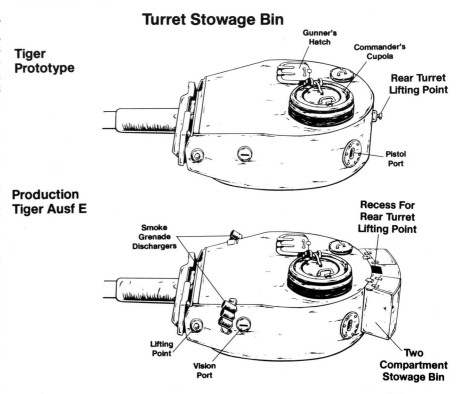

Tigers were introduced into new units for training during late 1942 and early 1943. These new production Tigers of sPzAbt 502 are engaged in summer training during 1943 and are equipped for tropical use with Feifel dust filters on the rear engine deck. These filters were later deleted from Tigers not intended for tropical service. The turret number is in Black.

'S' Mine Dischargers

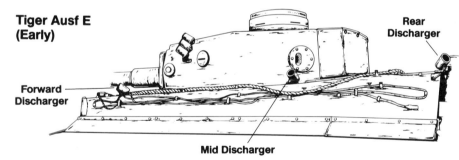

Tiger Ausf E (Early)

Rear Discharger

Forward Discharger

Mid Discharger

'S' Mine Discharger Unit, Three Per Side

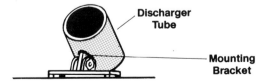

Discharger Tube

Mounting Bracket

The Feifel air filters were mounted on both the port and starboard sides of the rear engine deck above the mud flaps. The three short tubes on the hull top are "S" anti-personnel mine dischargers. The Feifel air cleaners were prone to damage after short periods of service and were often removed. They were, however, effective in preventing premature wear and damage to the engine from dust.

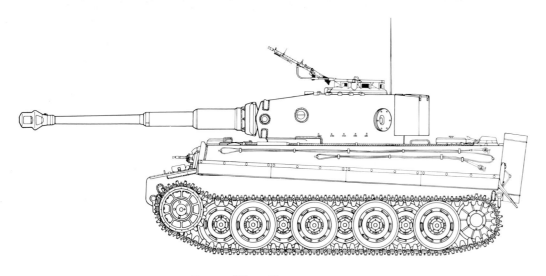

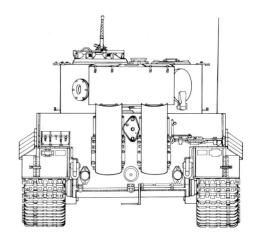

Specifications

Tiger ausf E

Length . . . . 27 feet 9 inches
Width . . . . 12 feet 3 inches
Height . . . . 9 feet 4 ⅔ inches
Weight . . . . 56 tons

Armament
  Main . . . . 88mm KwK 36 cannon
  Secondary . . . Two 7.92mm machine guns

Engine . . . . One 700 hp Maybach HL 230 P45 V-12 gasoline engine.

Performance
  Speed . . . . . . 23 mph
  Range . . . . 73 miles (road) 42 miles (cross country)
Crew . . . . Five

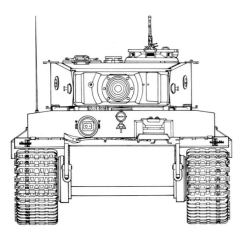

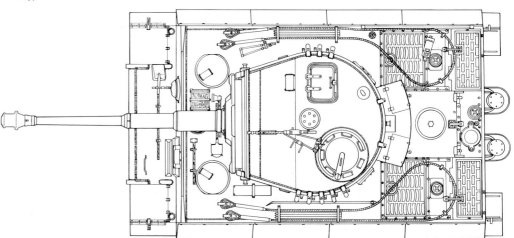

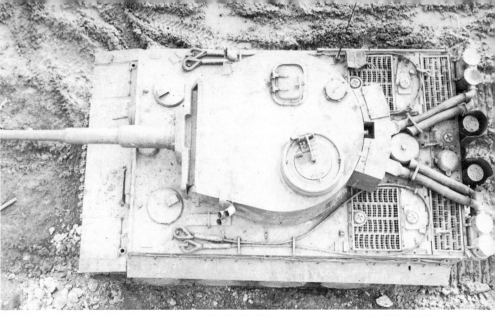

A German officer explains the next series of maneuvers to Tiger crews of sPzAbt 502 during Summer training. The circular object in front of the crewman is the turret roof fan. The recess in the turret stowage box was to clear the rear lifting hook used for removing the turret. The small pipe fitting on the cupola rim was a mounting for an auxiliary rangefinder.

This Tiger, 131 of sPzAbt 504, was captured in North Africa and evaluated by the British. The vehicle was in excellent condition when captured, with most of the internal and external stowage items in place. This Tiger has standard mudguards and headlights.

Turret Escape Hatch

Tiger Ausf E (Early)

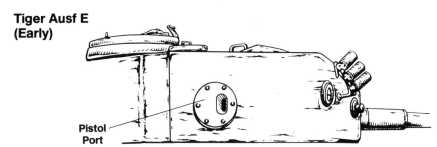

Pistol Port

Tiger Ausf E (Mid Production)

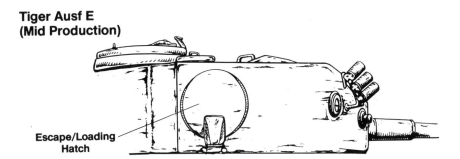

Escape/Loading Hatch

Tiger 131 was displayed by the British in Tunisia after being repainted for an inspection by King George VI. The shield on the mudguard is from the British First Army which captured this vehicle. The large round loading hatch on the turret rear replaced the second pistol port found on early production Tigers.

Ferdinand and *Elefant*

Although the Porsche VK.4501(P) tank prototype had lost the Tiger production competition to the Henschel VK.4501(H) prototype, ninety Porsche vehicles were ordered from the Steyr-Daimler factory at Nibelungen as a hedge against possible delays in production of the Henschel-built Tiger.

The VK.4501(P) prototype had been developed from Porsche's earlier VK.3001(P) project and used the same basic powertrain which consisted of two air-cooled gasoline engines driving electrical generators. These generators drove two electric motors which provided power to the tracks. In the event, the air-cooled engines proved to be a source of continual problems, while the electric motor drive system used an enormous amount of copper (which was a critical war material). These factors, as well as the superior handling and more familiar mechanical layout of the Henschel prototype, had resulted in Henschel receiving the Tiger production contract.

Because of the problems in the Porsche powertrain design, the German Army did not want to put the vehicle into service. The prototypes were relegated to the training role; however, some use had to be found for the ninety chassis in storage at Nibelungen. It was determined that these chassis were suitable for conversion into tank destroyers, mounting a long range, high powered anti-tank gun.

During 1942, eighty-five of the Porsche chassis were scheduled to be moved from Nibelungen to Alkett for conversion to the tank destroyer role. Dr. Porsche personally supervised the design work and the resulting vehicle was nicknamed *Ferdinand* (Porsche's first name) in his honor. The vehicle's official designation was *Sturmgeschuetz mit 8.8cm PaK43/2* (SdKfz 184), although it was better known as the *Jagdpanzer Ferdinand*.

While the design work for the conversion was being carried out, another demand was received from Hitler, ordering that the Ferdinands be built and sent to the front as quickly as possible. This resulted in the *Nibelungenwerke* at Nibelugen undertaking the conversion work on all the chassis, rather than sending them to Alkett.

The troublesome air-cooled engines of the VK.4501(P) were discarded and replaced with 300 hp Maybach HL120 water-cooled engines, while the electric drive system was retained. The engines were moved from the rear of the hull to the hull center so that the superstructure for the anti-tank gun could be set at the rear of the vehicle along with the track drive sprockets. The frontal armor on the hull was increased to 200MM with the addition of bolted-on armor plates, and the new superstructure had 200MM of frontal armor. Side armor was 80MM except for the lower hull which had 60MM armor. The rear armor was 80MM thick.

The main armament consisted of an 88MM Pak 43/2 L/71 cannon, which was a longer and more powerful version of the main gun used on the Tiger Ausf E. Originally there was no secondary machine gun armament installed for protection from enemy infantry.

The driver and radio operator sat at the front of the hull and the commander, gunner, and two loaders were located in the new superstructure. A large round escape hatch was fitted in the rear plate, with a smaller shell disposal hatch mounted inside the large hatch. Several vision slits and pistol ports were provided, but vision from inside the vehicle remained very poor. Additional hatches were installed in the roof, along with a ventilator and fan.

The *Ferdinands* were issued to *PzJagAbt* 653 and 654 in the Spring of 1943. They were committed to action during the great Kursk offensive near Orel in Russia. In combat they proved to be effective tank destroyers when engaging targets at long distances; however, their lack of secondary armament quickly became an urgent problem. Russian infantrymen were able to put several of the vehicles out of action by close-in infantry attacks and, at one point, *Ferdinand* crews were firing MG 42 machine guns down the barrels of

the main guns while the gunners searched out groups of Russian infantry with the main gun sights. Additionally, a number of Ferdinands were modified with crude platforms mounted at the rear of the vehicle to carry five *Panzergrenadiers* for protection from Russian infantry attacks. These men suffered very heavy losses and proved unsuccessful in halting further *Ferdinand* losses. The surviving vehicles remained in Russia for several months following the Kursk offensive before being sent back to Germany for repair and modification.

Several modifications were made to the *Ferdinands* that were returned to the *Nibelungenwerke*. A cupola derived from the StuG III ausf G self-propelled gun was installed on the superstructure roof, the tool stowage was revised, and a bow mounted MG 34 machine gun was added to provide the vehicle with a degree of self-defense against infantry attack. A total of forty-eight Ferdinands underwent the modification program and upon completion were redesignated as the *Panzerjaeger Elefant*.

These vehicles were re-issued to *PzJagAbt* 653, for use in Italy during early 1944. Once again they proved to be successful as long range anti-tank guns; however, in the hilly terrain of Italy they were even more ungainly than they had been in Russia. The suspension proved to be highly vulnerable to mines, and a number of vehicles were lost due to mechanical breakdowns. A number of these broken down vehicles were towed to critical strong points and dug in for use as static defense positions. As such, they proved to be very effective in stalling Allied armor and infantry advances. In the event, the lack of reliability and good maneuverability hampered the Germans in making the best use out of the excellent armament of the *Ferdinand/Elefant*.

This damaged *Ferdinand* self-propelled gun of PzJagAbt 654 was a survivor of the battle at Kursk during July of 1943. Returned to Germany for repairs, it retains Dark Yellow and Olive Green camouflage. Remanufactured *Ferdinands* were deployed to the Italian front, and redesignated as PzJag Elefant.

This *Elefant* was captured by American forces and returned to the United States for evaluation. Today it is on display at the Aberdeen Proving Grounds Museum in Maryland.

Ferdinand And *Elefant* Self-Propelled Guns

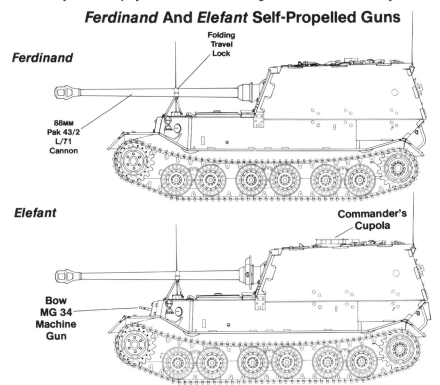

Ferdinand

Folding Travel Lock

88MM Pak 43/2 L/71 Cannon

Elefant

Commander's Cupola

Bow MG 34 Machine Gun

This *Elefant* of PzJagAbt 653 suffered the fate of most of these vehicles in Italy. Designed for open long-range anti-tank engagements, the vehicles were ill-suited for the hilly terrain and muddy conditions found on the Italian front. Many disabled vehicles were towed to fixed positions where their 88MM guns could command a critical road or other military objective.

This abandoned *Elefant* was put out of action when a mine damaged its suspension. Often the loss of a roadwheel resulted in the loss of the vehicle. This *Elefant* is camouflaged in Dark Yellow with a mottle of Olive Green and Red Brown and has a coating of zimmerit anti-magnetic mine paste on the lower sides.

Tigers in Service

The first German unit to be equipped with the Tiger ausf E was the 1st company of *sPzAbt* 502, which deployed with four Tigers to the Leningrad front during August of 1942. In the unit's first day of combat, all four of the Tigers were knocked out, although three were later recovered. The Tigers had not been used properly, being employed over terrain that was unsuited for tank action. The area was swampy and heavily forested which restricted the Tigers to established trails through the forest. Experienced Russian anti-tank gunners in well prepared and concealed ambush positions were able to quickly disable the Tigers, although none actually had their armor penetrated.

After repairs and reinforcements, the Tigers of 1st company went on to play a major role in the fighting on the Russia front, destroying one quarter of all the Russian tanks knocked out during the four months of the Russians' 1943 offensive. This accomplishment was no small feat, since the company had no more than seven Tigers available for action at any one time.

Other units were quickly trained and equipped with the Tiger and by the end of 1942, Tiger formations had been deployed to Russia, Africa, and Italy. Training centers were established in both Germany and France and eventually Tigers would be in service with ten Army heavy tank battalions, one training battalion, three SS heavy tank battalions, and the *Grossdeutschland* Panzer-Grenadier Division. A few additional formations received limited numbers of Tigers.

Tiger battalions were corps-level formations and could be used independently or to assist other units in breaking through enemy lines or holding back enemy counterattacks. In times of heavy losses or shortages of equipment, Tiger units were sometimes consolidated with other armor formations or formed into *Kampfgruppen* (battle groups), depending on the tactical situation.

Tiger Improvements

As production of the Tiger picked up in tempo and reports from combat units began coming in, it became obvious that improvements could be made in the performance and effectiveness of the Tiger. Kurt Arnoldt secretly kept continuing contacts with both officers and enlisted men in the Tiger battalions, even though this was a serious criminal offense. Officially, there was to be no contact between the troops using the equipment and the manufacturer. The first area to come under criticism was the reliability of the Tiger's chassis and powertrain.

Gradually the various problems reported were worked out, although some were never solved completely. The problems of ice and snow freezing on the interleaved roadwheels was not solved until introduction of the Tiger II with overlapping, not interleaved, roadwheels. The original 650 hp engine was replaced with a 700 hp engine and the transmission was improved, and with these improvements the Tiger performed reasonably well in normal use. Even with the more powerful engine, there was no reserve power or strength to tow other disabled Tigers without risking a break down. Only the existence of heavy tank recovery vehicles could correct this, and these vehicles were always in short supply since they were also intended to support Panther units.

Other improvements to the Tiger were actually efforts to standardize the German tank production program through the use of standardized components which would be used in a number of other German tanks. By this method the German Army sought to speed production and simplify their spare parts situation. During late 1943, Tigers began leaving the production line with a lower cast commander's cupola which had been designed for use on the Tiger II ausf B. The external "S" mine dischargers were deleted when the turret was modified during the Fall of 1943. The revised turret mounted the *Nahverteidigungswaffe* (close-in defense weapon), a traversable roof-mounted mortar that was loaded from inside the tank. It could fire smoke or anti-personnel shells which were similar to the old "S" mines in function. The turret ventilating fan was moved to the center of the turret roof to make room for the close-in defense weapon, and the turret-mounted NbK 39 smoke grenade launchers were also deleted.

Changes to the hull included deleting one of the Bosch front headlights, with the remaining one now being mounted in the center of the driver's front armor plate.

During early 1944, the Tiger's running gear was changed and new production Tigers were fitted with the resilient steel-rimmed roadwheels developed for use on the Tiger II and Panther II. Additionally, the new running gear had the outer (third) wheel on each axle eliminated, which helped solve the ice and mud buildup problem. The use of Feifel air cleaners had been discontinued in 1943, although they were often seen on older Tigers for some time after this. With the adaption of the cupola for the Tiger II ausf B, the Tiger ausf E also was fitted with the monocular TZF9c gunner's sight, which required only one sight hole in the gun mantlet.

Eighty-four Tiger ausf Es were converted to the command tank role with the addition of extra radio equipment and antennas. These vehicles had the ammunition stowage for the main gun reduced to sixty-six rounds, while ammunition for the MG 34 machine guns was reduced to 4,050 rounds. The command tank was designated *Panzerbefehlswagen mit 8.8CM KwK L/56*, or *Panzerbefehlswagen Tiger ausf E*.

Almost all late production Tiger ausf Es had factory applied Zimmerit anti-magnetic mine cement on all vertical surfaces, and most were built with the sheet metal shields around the exhaust stacks, although the style of these shields changed several times. In February of 1944, the Tiger's official designation was changed, now being designated the *Panzerkampfwagen Tiger ausf E*.

A Tiger of sPzAbt 501 leads a column of PzKpfw III medium tanks along a narrow road through a coastal town in Tunisia. Early Tiger battalions used PzKpw III tanks, armed with 75MM howitzers to support the Tigers. This Tiger has a dust cover fitted over the muzzle of the main gun to keep it clear of dirt and debris.

Camouflage and Markings

Because they served on many fronts, attached to a variety of other units, and in large multi-unit formations, Tiger battalions used more distinctive tactical markings, and carried a greater variety of these markings than most other German tank units.

The first Tigers issued to front line units during mid-1942 were delivered in overall Dark Gray (RAL 7027). In the Winter of 1942-43, washable White paint was used as camouflage in snow-covered areas. The Tigers of *sPzAbt* 501, which deployed to Africa during late 1942, were camouflaged in Desert Brown (RAL 8020) and while Dark Gray was authorized to be used as a second color in a disruptive camouflage pattern, there is no evidence that *sPzAbt* 501 ever painted their vehicles in this manner. In the more temperate climate of coastal Tunisia, many of the tanks of *sPzAbt* 501 were oversprayed with Olive Green (RAL 7008) to enhance their camouflage.

sPzAbt 504 had their Tigers camouflaged in overall Brown (RAL 8020) oversprayed with Olive Green (RAL 7008). It is not known if any Tigers went to North Africa painted in Dark Yellow (also known as Wehrmacht Olive), which was specified for use as an overall basecoat on all combat and front line support vehicles during 1943.

The camouflage colors used to paint the vehicles, with a wide variety of disruptive patterns, were Olive Green (RAL 7008 the light gray-green color first ordered for use in North Africa in 1941) and Red Brown (RAL 8017 which was more of a brown than a red). Tigers used all these colors in a wide variety of schemes and applications.

During August of 1944, to reflect the needs of a changing war situation, the German Army added new camouflage colors which were intended to be used in place of the older shades. A new Olive Green (RAL 6003) was introduced, along with a new Red Brown (RAL 8012). The new Red Brown was more red than the older Red Brown, while the new Olive Green was somewhat darker than RAL 7008, and was often used as a primer color on many vehicles beginning in November of 1944.

Mud Guards

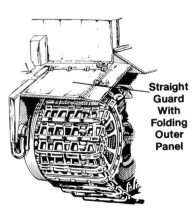

**Tiger Ausf E
Standard Mud Guard**

Straight Guard With Folding Outer Panel

**sPzAbt 501
Modified Mud Guard**

Thinner Narrow Guard Without Folding Section

Tiger 142 of sPzAbt 501 advances down a road in Tunisia. The Tiger has the unit field modifications, the lowered headlights and modified mudguards, that made the tanks assigned to sPzAbt 501 unique. The tank is camouflaged in overall Brown, possibly with an over spray of Olive Green. The turret number is Red outlined in White outline.

The very dirty and weathered appearance of this Tiger contrasts sharply with the newly painted Red and White turret numbers. Dust covers were used over the guns to reduce maintenance problems caused by dust contamination. Dust was the greatest problem in Africa, followed by the wear on the engine caused by the extreme heat.

In the last months of the war, Dark Gray was also used on a number of vehicles, both as a primer and as a camouflage color. It must be noted that older paints were almost always used until supplies were exhausted, so many older vehicles carried new paint colors while newer vehicles often appeared in older colors. It should also be noted that many German manufacturers used Red Oxide primers extensively, and some of these primer paints appeared on new vehicles.

The markings used on Tigers were perhaps more varied than those of any other German combat vehicle. As the Tiger battalions moved from engagement to engagement, from one command structure to another, they came under the command of many different formations. This led, in many cases, to the Tiger units adopting different markings and even marking systems, especially in the tank identification numbers. Most Tiger units used the standard German Army three-digit system of vehicle identification: the first digit denoted the company, the second digit denoted the platoon, and the third digit denoted the individual vehicle within the platoon. In some Tiger battalions, only the company number was used to identify the vehicle, in others, only the platoon and individual vehicle number, while other units used only the vehicle number.

In addition, a wide variety of number styles and colors were found in Tiger units. Many Tiger battalions used fairly consistent numbering; others changed not only styles, but also systems. Some of these resulted from being attached to another unit, and as a result having to renumber the Tigers. In other cases, these changes appear to have resulted from changes in commanding officers, a new CO changing things to suit his own preferences.

This Tiger of sPzAbt 501 during late 1942 reveals the lowered position of the headlights, which was a sPzAbt 501 field modification. The unit relocated the lights to move them from their exposed location on the upper corners of the hull where they were easily damaged. The modified front mudguards, which were fitted to all sPzAbt 501 Tigers, also distinguished 501's Tigers from those of its sister unit sPzAbt 504.

Organization

The theoretical level of equipment for a Tiger heavy tank battalion consisted of a headquarters company and four companies, each with a headquarters section and three platoons of four Tigers each. This organization gave a Tiger battalion total strength of fifty-nine Tigers. In reality, very few battalions were able to field maximum strength and the organization was later revised to a total of forty-five Tigers in a battalion. The revised organization had a headquarters platoon (instead of company) with three Tigers, and three companies, each with two headquarters Tigers and three four-tank platoons. Even this level was difficult to maintain and most Tiger formations remained understrength throughout the war.

The high level of preventive maintenance needed to keep the Tiger running required the best in crews and repair troops. As the war progressed and casualties were incurred in many of the early, well-trained Tiger crews, maintenance suffered. The German training establishments had a difficult time in keeping up the level of crew training to the earlier standards.

As Allied tanks improved, the weight and relative lack of maneuverability of the Tiger became a pressing problem, although the shift to defensive fighting put the Tiger into its best operating environment. The Tiger remained one of the most formidable weapons of the war, right up to the end. During 1945, Tigers and Tiger IIs were simply overwhelmed by huge numbers of Allied tanks, the air superiority of the Allied air forces, and by shortages in fuel, ammunition, and spare parts. When, in the Spring of 1945, the U.S. Army shipped an upgunned and uparmored M26 Pershing (nicknamed the Tiger Killer) to Europe, it never encountered a Tiger or Tiger II in combat. By April of 1945, the Tiger was nearly extinct in Europe.

Headlights

Tiger Ausf E Standard

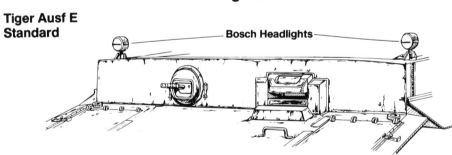

sPzAbt 501 Field Modification

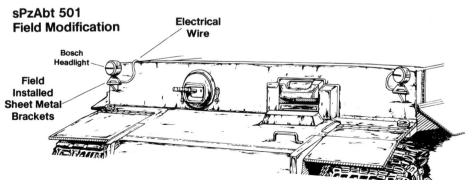

An early Tiger ausf E of sPzAbt 501 advances down a road in Tunisia. This Tiger clearly reveals the field modifications made to sPzAbt 501 Tigers, the relocated headlights and cut down fenders.

Tiger 142 of sPzAbt 504 advances through the brush in Tunisia. The tank is overall Brown with the Dark Red turret numbers almost completely hidden under the layers of dust. The small White rhomboid on the corner of the hull denotes the 1st company. It was not unusual for crews to travel with the tank's hatches open to lessen the heat build-up inside the vehicles.

This Tiger of sPzAbt 501 has been assigned to PzRgt 7, 10.PzDiv where the Tigers were organized as the 7th and 8th companies of the regiment. This tank has seen a lot of action and is missing the side skirts and headlights, and has had the front mud guards damaged. The light spot on the gun mantlet is where an enemy shell partly penetrated the mantlet.

A Tiger crewman visually inspects the barrel of the main gun. After cleaning, the bore was checked for cleanliness and damage by shining a bright light into the chamber and down the barrel. This made it much easier to see the bore clearly and spot any problem areas.

This Tiger of sPzAbt 503 is overall Dark Gray with a coat of White winter camouflage paint applied directly over the original camouflage. The winter camouflage paints were scarce and were applied in the field resulting in a very rough appearance. The turret numbers are White with a thin Black outline. This Tiger also carries a PzKpfw III stowage bin.

The crew of a Tiger of sPzAbt 503 loads ammunition through the turret which shows the scars of hits from an anti-tank rifle and an anti-tank gun. German armor plate was of high quality, and although many Tigers were destroyed by Allied weapons, the Tiger's armor protection was among the best in the world.

sPzAbt 503 received its first Tigers during late 1942 which were delivered in overall Dark Gray. This Tiger of the 2nd company is being overhauled in the Rostov area during early 1943 and has a PzKpfw III stowage bin fitted to the turret rear. Most Tiger units on the Russian front were forced to maintain their vehicles under very poor conditions.

The crew of this Tiger cleans out the barrel of the 88mm KwK36 main gun. Two or three men were required to swab out the bore, and one crewman remained in the turret to clean out the chamber and check the bore. This Tiger of sPzAbt 502 was later destroyed during August of 1944 by a Russian 85mm anti-tank gun.

Gun Mantlet

**Tiger Ausf E
(Early Production)**

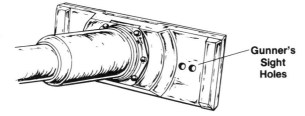

Gunner's
Sight
Holes

**Tiger Ausf E
(Mid Production)**

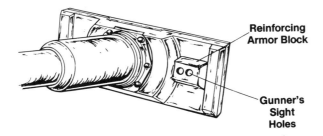

Reinforcing
Armor Block

Gunner's
Sight
Holes

Any major work that had to be done on the transmission or final drive usually required the removal of the turret. Repair crews position lift hooks and cables from an overhead gantry crane to the lift points on the Tiger's turret prior to lifting the turret off the hull.

Transporting Tigers by rail usually involved removing the wide battle tracks and installing the narrower transport tracks. This Tiger, however, has been put onto a flatcar with the battle tracks installed. Although the narrow clearances in Germany and Europe led to the use of narrow transport tracks, in Eastern Europe and in rural areas the railroads were often open enough to allow the battle tracks to be left on.

A real problem with the interleaved roadwheel suspension of the Tiger was the build up of mud. If the build up became too great, it could jam the wheels and lead to thrown tracks or jammed sprockets. Frozen mud could leave a Tiger completely immobile and, on occasion, led to the loss of the vehicle.

One of the reasons Tiger crews had such confidence in their vehicles was the Tiger's ability to absorb punishment. This Tiger driver examines a direct hit on his visor shield. The force of the hit has sprung the upper visor segment; however, the driver was uninjured. Most Allied tank guns could not penetrate its front armor at ranges beyond the killing range of the Tiger's main gun.

A crewman reaches through the large engine access hatch of a Tiger of sPzAbt 503 during the Spring of 1944 to work on the engine. The original Black outlined White turret numbers have been filled in with Black.

Even in the field, Tiger crews and repair troops were able to strip the vehicles down and repair a wide range of mechanical problems or damage. German recovery and repair companies rebuilt and saved hundreds of tanks and other vehicles that otherwise would have been lost to enemy action.

The crew of this Tiger has raised the rear engine cover to gain access to the engine compartment. Normally, repairs were made in rear areas and under cover to protect the immobile Tiger from detection and destruction by enemy aircraft.

The shape of the crater left by an anti-tank projectile indicates a fairly straight hit on the side of this Tiger. The shell tore through the sheet metal side skirt but did not penetrate the side armor. Like most tanks, the Tiger was more vulnerable to the sides and rear where the armor was not as thick as the front.

The crew of this Tiger of sPzAbt 503 are loading ammunition for both the 88MM main gun and the bow and coaxial MG 34 machine guns. The tank shows the effects of continued use with many of the smaller fittings removed. In action, many of these fittings were actually broken off and lost.

During training and maneuvers, Tiger crews sometimes gave demonstrations of their vehicles to civilians. This Tiger, camouflaged with brush, is engaged in just this sort of demonstration during the Summer of 1943. The vehicle is in excellent condition with all mudguards and stowage intact and undamaged, and probably has never seen action.

The lack of dedicated heavy tank recovery vehicles was a real problem for Tiger units. On occasion, Tigers broke down while attempting to tow other disabled tanks. Normally it required three eighteen ton SdKfz 9 tractors to tow one Tiger. This Tiger of sPzAbt 503 has had the front roadwheel removed to prevent the track from being jammed by debris, a fairly common practice.

24

This Tiger ausf E of sPzAbt 504 fought in Tunisia during 1943.

A Tiger ausf E of the 8th *Kompanie*, *Das Reich* PzDiv in Russia during April of 1943.

Tiger ausf E A02 was attached to the 9th *Kompanie*, *Grossdeutschland* Division on the Russian Front during 1943.

This mid-production Tiger ausf E served with sPzAbt 503 against the Russians during the Summer of 1943.

A Tiger ausf E of SPzAbt 508 on the Italian Front during 1944.

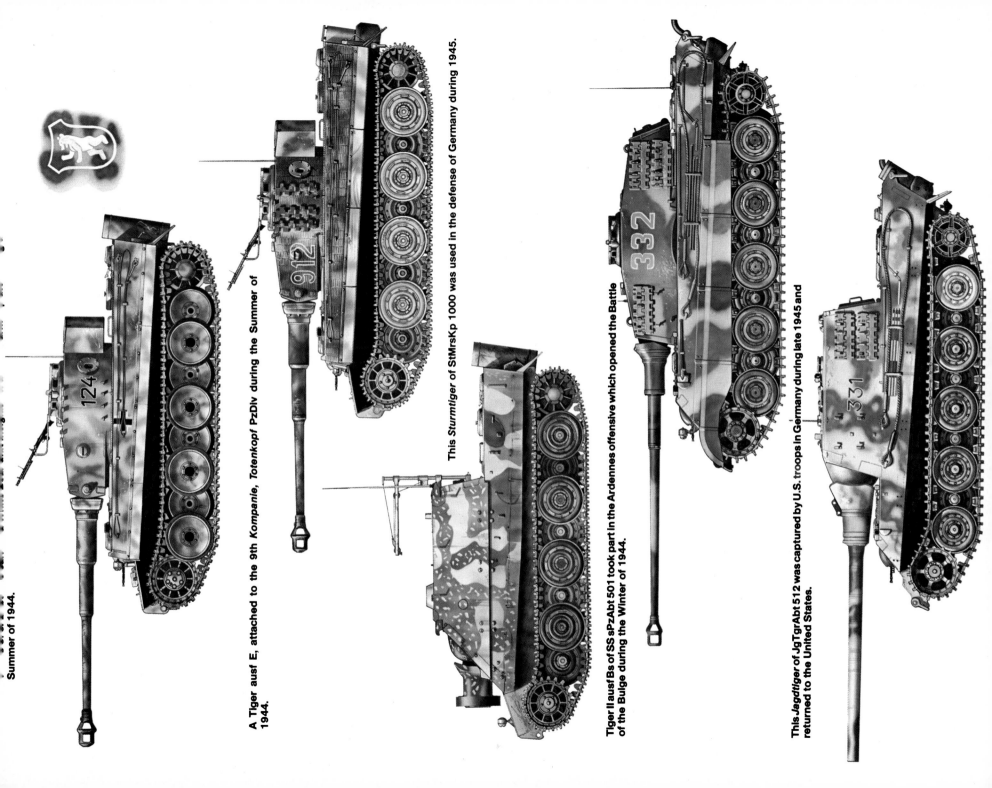

Summer of 1944.

A Tiger ausf E, attached to the 9th *Kompanie, Totenkopf* PzDiv during the Summer of 1944.

This *Sturmtiger* of StMrsKp 1000 was used in the defense of Germany during 1945.

Tiger II ausf Bs of SS sPzAbt 501 took part in the Ardennes offensive which opened the Battle of the Bulge during the Winter of 1944.

This *Jagdtiger* of JgTgrAbt 512 was captured by U.S. troops in Germany during late 1945 and returned to the United States.

Tigers of SS PzDiv *Das Reich*, move down a road in the Perestschnaja area of the Eastern front. The vehicles are camouflaged in Dark Gray with over sprayed areas of Dark Yellow and Olive Green and the turret numbers are in a White outline style. By this time, many Tigers no longer carried the sheet metal heat shields around the exhaust stacks.

This Tiger (123) of sPzAbt 503 during the Summer of 1943 reveals the strengthened mantlet and the thickened reinforcing bar around the gunner's sight holes. This modification was introduced on the Tiger production line as a result of lessons learned during combat.

This Tiger of sPzAbt 503 reveals the kind of damage commonly suffered by the mudguards and skirts during normal combat operations. The vehicle's armor carries scars from anti-tank rifle and armor piercing bullet strikes, and the roadwheels have shrapnel damage. The turret numbers are Black outlined with White.

The crewmen of this Tiger of sPzAbt 503 are replacing the rear idler wheel. Although of a different style, the turret numbers on this Tiger are also in Black outlined in White. This is a replacement vehicle with new numbers, which accounts for the difference in appearance.

27

The Tiger was introduced into the African campaign during late 1942, when sPzAbt 501 was sent to reinforce the German forces in Tunisia. This tank of the 1st company, sPzAbt 501 is camouflaged in overall Brown and has the turret numbers in Red outlined in White. This vehicle is an early production Tiger with a pistol port on the starboard side of the turret.

The White outline turret number on the turret rear of this Tiger of sPzAbt 508 denotes the 2nd company. The water jerrican is Black with a White cross to rapidly identify it as water. There is a solid mud buildup between the last two roadwheels on the side with the thrown track. Keeping the suspension free of excess mud was a never ending problem with the interleaved wheels on the Tiger.

This Tiger of sPzAbt 501, was attached to PzRgt 7, 10.PzDiv, and has large Red and White turret numbers applied over an overall Brown camouflage. This vehicle is fitted with early style square heat shields on the rear exhaust manifolds and stacks. These shields were later changed to a rounded style with cooling vents.

Exhaust Cooling Shrouds

Tiger Ausf E

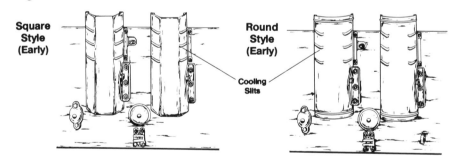

Square Style (Early)

Round Style (Early)

Cooling Slits

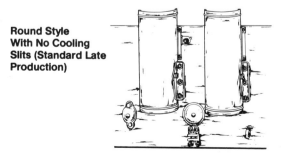

Round Style With No Cooling Slits (Standard Late Production)

Officers confer on top of a Tiger of the *Das Reich*, SS PzDiv during early 1943. The vehicle carrys a camouflage scheme of Dark Gray, over painted with Dark Yellow and Olive Green. By early 1943, the old Dark Gray base color was ordered to be covered or supplemented by newer colors (Dark Yellow, Olive Green, and Red Brown) on all front line combat and support vehicles.

The crew of this Tiger of sPzAbt 501 loads ammunition for the main gun through the loader's hatch. The loader was responsible for placing the shells in the internal stowage bins in the order he wanted them stowed. Since the Tiger used both High Explosive (HE) and Armor Piercing (AP) ammunition, the loader had to know exactly where the different types of shells were stored.

Tiger S13 carries the markings adopted by *Das Reich* for the Kursk offensive during July of 1943. The division sign on the front armor is in White, as are the outline turret numbers. The demon marking on the turret in White was the mascot of the Tiger (8th) company of the division.

This Tiger ausf E of sPzAbt 501 during early 1944, reveals the major external changes to mid-production Tigers. The drum cupola has been replaced with the lower cast cupola developed for the Tiger II/Panther II, and the two headlights have been replaced by a single headlight mounted in the center of the front plate. The cupola hatch is mounted on a vertical pivot shaft, opening upward and turning to the port side of the turret.

These Tigers of the 8th company of SS PzDiv, *Das Reich* during April of 1943, reveal the Dark Yellow used as a second camouflage color over the original Dark Gray base color. These newly issued vehicles also reveal the early 1943 configuration of the Tiger ausf E; Feifel air cleaners and hull mine dischargers still being carried, with stowage fittings for spare tracks added on the turret sides.

This late production Tiger ausf E of sPzAbt 503, advancing through a French town in Normandy during August of 1944, reveals the new roadwheels and monocular gunner's sight. With the new road wheel design, the outer row of wheels, normally removed during rail transport, have been eliminated entirely. This Tiger is Dark Yellow with heavy mottling of Olive Green and Red Brown with the turret numbers in Black with White outlines.

Somewhat battered, these Tigers of sPzAbt 508 move up a road in Italy during 1944. The Tiger was capable of good speed on hard level roads and, although many considered it to be somewhat underpowered, the Tiger suffered relatively few engine failures. Engine life was shortened considerably, however, if the tank was used to tow other tanks.

Commander's Cupola

**Tiger Ausf E
(Early)
Drum Style Cupola**

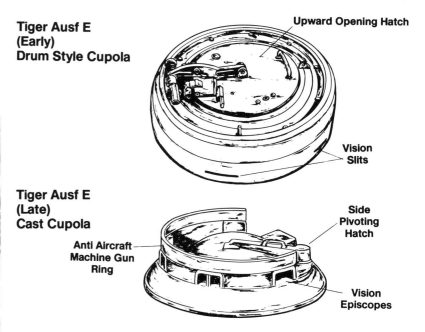

Upward Opening Hatch

Vision Slits

**Tiger Ausf E
(Late)
Cast Cupola**

Anti Aircraft Machine Gun Ring

Side Pivoting Hatch

Vision Episcopes

Tigers, along with many other German tanks, were coated with zimmerit anti-magnetic cement applied to the vertical surfaces as protection from magnetic mines. Normally the zimmerit coating for the Tiger was applied at the factory with a toothed trowel or blade which produced a characteristic rough texture.

Tiger 201 of sPzAbt 501, hidden in a Russian village during early 1944, is being used as a long range anti-tank gun. The Tiger was intended to be an offensive tank; however, it proved to be better suited for the defensive role with its powerful gun and excellent sights. Tiger 201 is camouflaged with a partial coating of white wash under a heavy layer of mud.

This Tiger of sPzAbt 501 carries extra track sections on the turret roof as it advances through a Russian forest. Mid production Tigers standardized the single headlight configuration along with having many small fittings removed to simplify production. The turret numbers are Black outlines filled with Yellow Ochre.

Two sPzAbt 501 Tigers advance along a road in Poland during 1944. Tigers rarely operated alone and standard tactics stressed mutual support. During 1944, there were often shortages of camouflage paints, and supplies had to be stretched. Since tanks usually fired from hull down positions, where only the turret was visible, winter whitewash was usually applied to the turrets but not always to the hulls.

31

A new production Tiger ausf E is loaded onto a train still fitted with its wide battle tracks. If at all possible, tanks were shipped without changing the tracks to save time. This tank has a factory applied White winter camouflage and no unit markings. Receiving units were responsible for applying the vehicle's tactical markings as required.

This mid-production Tiger is fully coated with zimmerit anti-mine cement. The open driver's and radio operator's hatches reveal the periscopes fitted to the hatch. This Tiger has also had the first outer road wheel removed to prevent jamming or obstruction of the track and sprocket.

This Tiger is chained down and has the tracks blocked, ready for rail shipment to the front. It is fitted with the narrow transport tracks, with the battle tracks stowed under the vehicle. Because of the wear and tear on tank engines when they were driven long distances, most long distance tank transport was done either by rail or heavy trailer.

This late production Tiger ausf E reveals the major changes found on vehicles produced after February of 1944. The gunner's sight has been changed to a monocular sight, requiring only one hole in the mantlet and the roadwheels are the later steel rimmed resilient type also used on the Tiger II and on a short production run of Panther ausf Gs.

A mid-production Tiger ausf E of sPzAbt 502 crosses a ditch in the Kurland sector of the northern Russian Front during early 1945. This Whitewashed vehicle has Red turret numbers. This Tiger has the first set of roadwheels removed and also has the first outer wheel damaged.

This Tiger of sPzAbt 508 is Dark Yellow over sprayed with Olive Green and Red Brown. The vehicle has had the bow MG 34 machine gun removed, indicating that it is well behind the lines, traveling to a unit organization point. Although rail transport was preferable, tanks were driven fairly long distances when rail travel was not possible.

A Tiger of the 1st company, SSsPzAbt 101 drives through a French village past a *Schwimmwagen* amphibious car during 1944. Tigers of 1st Company used White outlines on the turret numbers (if the number was painted on a dark area of the turret). Tiger 133 reveals this practice with White outline numbers against the dark camouflage on the turret side. The outlines were often filled in with brushed paint or masked off during spraying.

Pulling the barrel and breech of the 88mm KwK36 main gun into position required the efforts of most of the crew. The shiny threaded area on the end of the barrel is for attaching the muzzle brake to the gun tube. Timber sections are used to support the turret on the oil drums to prevent damage to the turret ring and gears.

Gun Mantlet

Tiger Ausf (Mid Production)

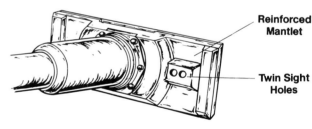

Reinforced Mantlet

Twin Sight Holes

Tiger Ausf E (Late Production)

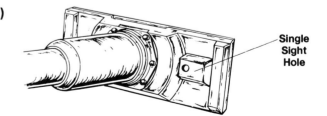

Single Sight Hole

In very soft ground or mud, Tigers were prone to digging in up to the mudguards. This vehicle was eventually pulled out of the mud, but only after a great effort. The tank is painted in Dark Yellow with a light mottle of Olive Green, with the upper heat shields are in Red Brown.

(Below) This mid-production Tiger of sPzAbt 508 on the Italy front during 1944 has lost a track. The vehicle is Dark Yellow with a heavy over spray of Olive Green. Many of the sPzAbt 508 Tigers were almost completely over painted with Olive Green. The turret number is a White outline. Tigers belonging to sPzAbt 508 often carried only the company number.

(Above) The high fuel consumption and lack of suitable cross-country transport led many Tiger units to carry extra fuel in drums on their tanks when traveling to a new area. These Tigers of sPzAbt 505 on the Russian Front during early 1944 have Black numbers which are partially obscured under the roughly applied Whitewash winter camouflage.

The Tigers of sPzAbt 506 carried a large "W" symbol on the turret rear, with a tiger painted over the letter in Black and Yellow. The color of the W denoted the company; the Tiger to the right has a White W (1st company) and tank in the foreground has a Red W (2nd company). 3rd company tanks carried a Blue W.

This Tiger of *Grossdeutschland* is being prepared for rail transport. The cable attached to the narrow transport track leads up to the drive sprocket which will pull the track up to the front of the tank where the crew will join the two ends. This Tiger is Dark Yellow with Red Brown patches and light Olive Green mottling with the turret number, A32, in Black.

Road Wheels

In the spring of 1944, sPzAbt 505 adopted a charging Knight as its unit symbol, retaining it to the end of the war. This Tiger is mostly Dark Yellow, with a Red Brown gun barrel. The Knight marking on the turret is Black and Red, and the number on the gun recuperator housing is in Black.

Tiger Ausf E (Early)

Rubber Rimmed Roadwheel

Tiger Ausf E (Late)

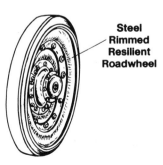

Steel Rimmed Resilient Roadwheel

Two eighteen ton SdKfz 9 tractors tow a disabled Tiger of sPzAbt 508 into a grove of trees for repairs. Between 23 and 25 May 1944, the 3rd company of sPzAbt 508 lost fifteen out of its sixteen Tigers, most to mechanical breakdowns. The lack of heavy recovery vehicles cost German units a number of vehicles they could have otherwise retrieved.

Tigers of *Grossdeutschland* arrive at a unit assembly point in Lithuania during August of 1944. These vehicles are overall Dark Yellow with a light pattern of Olive Green. The narrow transport tracks were placed back on the flatcars once the wide battle tracks were installed on the tanks. The side skirts are stored under the tanks and the outside roadwheels are in front of the tanks on the flatcars.

Crewmen and repair troops of *Grossdeutschland* prepare to install an engine in Tiger A31 during the Fall of 1944. The engine crane is mounted on a four and a half ton heavy truck. The vehicle number A31 is in Black and the Tiger is Dark Yellow with brushed patterns in Olive Green.

Tiger A31 has a full coating of zimmerit anti-mine cement and is unusual in that it retains the hull-mounted S anti-personnel mine dischargers on the corners of the rear hull. The dark patterns on the hull rear are a combination of wet mud and soot from the engine exhaust.

This burned out Tiger of the 1st company of SSsPzAbt 101 was destroyed in Villers-Bocage, France, on 13 June 1944. The White rhomboid marking of the 1st Company is about all that remains of the tanks original markings. The vehicle suffered an internal explosion, blowing the cupola hatch off the turret.

A Tiger of SSsPzAbt 101 advances through a French town during 1944. Although the turret numbers on this Tiger appear to be Red, 1st Company vehicles did not use Red for their iden- tification numbers. The markings are actually White outlines filled in with Olive Green. The Tiger is camouflaged in Dark Yellow with heavy over sprays of Olive Green and Red Brown.

The 3rd Company of SSsPzAbt 101 used Medium Blue turret numbers, usually outlined in Yellow, which are faintly visible on the turret side of this Tiger during the Spring of 1944. This mid-production vehicle has its single headlight moved to the old hull top position, and is equipped with older style roadwheels and a binocular gunner's sight.

Headlights

Tiger Ausf E (Early)

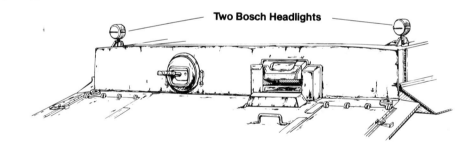

Two Bosch Headlights

Tiger Ausf E (Late)

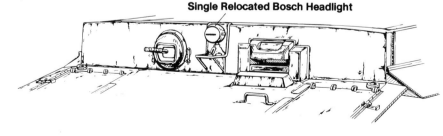

Single Relocated Bosch Headlight

Sturmtiger

The *Sturmtiger* assault weapon originated from a requirement issued by the German Army for a self-propelled heavy assault howitzer capable of destroying heavily fortified positions and large buildings. The weapon was envisioned to be used in urban warfare, like that encountered at Stalingrad. Originally the army requested the vehicle be armed with a 210MM howitzer; however, no suitable weapon was available. As a substitute, the Army chose the *Raketenwerfer* 61 L/54, a 380MM rocket launcher which had been developed for the German Navy as an anti-submarine weapon. It was decided that the Tiger ausf E chassis was the most suitable chassis for this new weapon system, and Alkett began development of the vehicle during 1943. In October of 1943, a prototype was successfully demonstrated to the Army and the vehicle was ordered into production during mid-1944. Brandenburger Eisenwerke built the superstructures and Alkett did the final assembly work on the vehicles in its Berlin-Spandau plant.

The superstructure for the new vehicle was massive with 150MM thick front armor and 80MM side and rear armor. The hull had 100MM front armor, 60MM at the sides, 80MM at the rear, while the roof was 40MM thick. A driver's armored visor and ball-mounted MG34 were placed in the front plate, along with a sighting aperture for the gunner's PaK ZF3X8 sight. Pistol ports were installed in the side plates, and a large round hatch was centered in the rear superstructure plate. A crane for loading the rocket ammunition was mounted on the starboard rear corner of the superstructure. The roof contained a ventilator and fan, a round hatch, and a two-part rectangular hatch for loading ammunition. The rear part of this hatch was spring balanced and contained a close-in defense weapon.

Inside the superstructure were six stowage racks for a total of twelve rocket projectiles. Each projectile weighed 345 kg (761 pounds) and was 1420MM (56 inches) long. The bore of the launcher was 380MM (15 inches) in diameter and the rockets had a maximum range of 5,670 meters (6,200 yards). A removable loading tray with rollers allowed the crew to load the projectiles manually into the projector when the barrel was set at 0 degrees elevation. A winch was mounted under the roof which enabled the crew to load and unload all the stowage racks and position the projectiles in the loading tray.

The rocket projectiles were thin-walled cases with multiple venturi holes in the rear of each projectile which allowed the exhaust gases to propel the projectile out of the projector. The projector barrel was vented and dispersed the rocket blast out through the front of the barrel. A spine at the rear of the projectile engaged the rifling in the bore, spinning the projectile as it went down the barrel. The breech was sealed by a breech block during firing. The crew consisted of five men: commander, driver, observer, and two loaders. The vehicle was officially designated the *38CM RW61 auf Sturmmoerser Tiger* but was more commonly known to the troops as the *Sturmtiger*.

Although the rocket projectile was effective in demolishing most targets, the *Sturmtiger* was slow and mechanically unreliable. Most were captured, usually after being abandoned by their crews after mechanical problems. Even with its heavy armor, the *Sturmtiger* was such a large target that a number were shot up and destroyed by Allied tanks. The Allied vehicles circled around to attack the *Sturmtiger* from the rear, where the armor was most vulnerable to attack.

The only major variant of the Tiger ausf E to reach production was the *Sturmtiger* built on a late Tiger chassis with resilient steel rimmed roadwheels. With 150MM of front armor, the *Sturmtiger* was virtually impossible to penetrate with any Allied tank ammunition. The vehicle was intended to destroy fortified positions and buildings with its huge weapon. (U.S. Army)

The rear of the *Sturmtiger* was its most vulnerable area and this vehicle had several AP rounds fired into its engine compartment, destroying the vehicle. The raised roof loading hatch is configured with a close-in defense weapon which replaces the old external S anti-personnel mine dischargers. The crane on the rear superstructure wall was used to load ammunition. (U.S. Army)

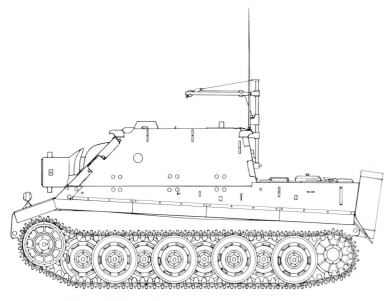

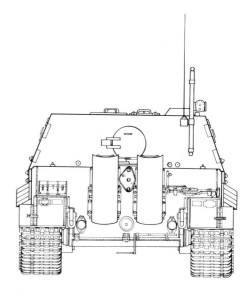

Specifications

Sturmtiger

Length . . . . 20 feet 8½ inches
Width . . . . 12 feet 3 inches
Height . . . . 11 feet 4 inches
Weight . . . . 70 tons

Armament
  Main . . . . 380мм Raketenwerfer 61 L/54 rocket projector
  Secondary . . . One 7.92мм machine gun.

Engine . . . . One 700 hp Maybach HL 230 P45 gasoline engine.

Performance
  Speed . . . . . . 25 mph
  Range . . . . . . 87 miles (road) 55 miles (cross country)
Crew . . . . Seven

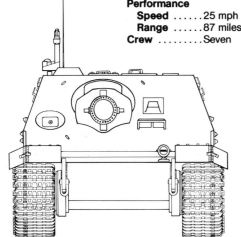

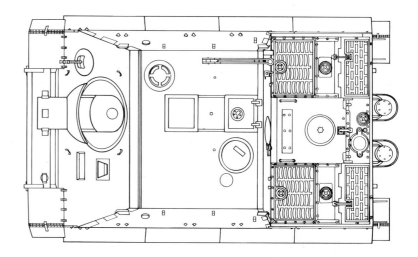

Tiger II Ausf B

The Tiger II was developed in answer to a requirement from the *Waffenamt* (weapons development section) of the German Army for a new tank to replace the Tiger ausf E. This requirement was issued because of continuing mechanical problems with the Tiger ausf E and because it was felt that there was a need to mount a larger weapon than could be accommodated in the Tiger ausf E.

Both Porsche and Henschel were asked to submit designs for the new heavy tank. Porsche designed two variants, both mounting the same turret and 88MM KwK43 L/71 main gun. The Type 180 had the turret forward in the center of the hull while the Type 181 had the turret mounted at the hull rear. Both vehicles were to use the gasoline-electric drive of the earlier VK.4501(P). The VK.4502(P) designs were again rejected in favor of the Henschel VK.4503(H), primarily because the hybrid gasoline-electric drive was still not as reliable as the conventional powertrain used in the Henschel vehicle. Additionally, the shortage of copper for the electric drive was now even more acute than it had been a year before.

The successful Henschel VK.4503(H) prototype was a conventional tank laid out in much the same manner as the earlier Tiger ausf E. The specification had required the best ballistic shape possible, and the VK.4503(H) had well-sloped armor all around. The front armor was increased to 150MM; however, the side armor remained 80MM. The project was given the highest priority, but it was delayed several months in order to collaborate with MAN, which was developing the Panther II tank.

The coordination led to adopting a great number of fittings as standard between both vehicles. This policy reduced production costs and simplified the spare parts supply situation. As a result, although the Tiger II was a development of the Tiger, it had the appearance of an enlarged Panther, and shared a number of parts with the Panther II. These common features included the basic engine deck details, cupola, hull hatches, resilient roadwheels, and numerous internal mechanical components.

During October of 1943, the final Tiger II mockup was inspected by Hitler and other high ranking German officials and in January of 1944, the first production vehicles began leaving the Henschel factory. The first fifty production Tiger IIs used turrets designed for the unsuccessful Porsche entry. These turrets had a curved front plate and mantlet which formed a serious shot trap below the mantlet (a shot trap is an area that deflects incoming enemy rounds into the vehicle, in this case, down into the driver's position). As a result, Henschel was requested to design a new turret. The Henschel design was not only more spacious and far better protected, it was also cheaper to manufacture. The Henschel turret was selected as standard and was installed on all subsequent production vehicles.

The Tiger II ausf B had an improved version of the Tiger ausf E powertrain, and was almost as fast on level ground; however, when it was called to tow other disabled tanks, the strain often caused the engine to fail. While Henschel and their subcontractors had made a number of improvements and the vehicle was better mechanically than the earlier Tiger ausf E, it was, unfortunately, the heaviest tank to serve during the Second World War. The Tiger II's weight and size caused the Germans numerous problems in both transporting and deploying Tiger II units. While easier to manufacture than the Tiger ausf E, the Tiger II ausf B still consumed huge quantities of steel and production manhours, resources that probably could have been more wisely used in producing greater numbers of a medium tank like the Panther.

The basic layout of the Tiger II ausf B was identical to that of the Tiger ausf E. The sloped glacis forced the elimination of the driver's visor which was replaced with a rotat-

This early Tiger II ausf B is fitted with the Porsche designed turret which used a binocular gunner's sight. The curved front plate created a serious shot trap which deflected incoming rounds down into the driver's compartment. The bulge for the cupola also was a weak area in the turret side armor. The production Henschel turret eliminated these deficiencies.

ing periscope mounted in front of the driver. The radio operator was provided with a fixed periscope mounted over the bow MG 34 machine gun mount. The two hull hatches lifted and rotated, as did the hatch on the turret cupola. The turret did not have a basket and was fitted with a large rear bustle which held twenty-two rounds of ready ammunition. Power traverse for the turret was the same as on the Tiger, and also suffered from a slow turning speed. The turret had been extended outward both in the front and in the rear to handle the weight of the 88MM KwK43 L/71 gun without overloading the turret ring mechanism, which was 1855MM (73 inches) in diameter.

The Porsche turret's front armor was 110MM, with 80MM sides and rear. The Porsche turret was fairly complex, with a very thick curved front piece and a bulge in the port side for the cupola. The later Henschel turret had a flat front plate of 180MM thickness, with 80MM sides and rear. Both turrets were equipped with a rear hatch and a loader's hatch in the roof. The roof on both turrets was 40MM thick. The Henschel turret had a conical *Saukopf* (boar's head) mantlet, which provided excellent protection for the gun mount.

The main armament of the Tiger II was the 88MM KwK43 L/71 cannon, which was widely considered to be the best all-round tank gun of the Second World War. It had a flat trajectory and the hitting power to destroy almost any Allied tank at extremely long range. The turret traverse, however, was extremely slow. The fastest time for a full 360 degree turn was nineteen seconds, although it often took as long as seventy-five seconds. Manual traverse usually required the services of both the gunner and the loader. It took 700 turns of the gunner's handwheel for one complete revolution of the turret, while the loader's handwheel required 680 turns for a complete traverse. The power traverse was certainly preferred, even though it was slow.

Secondary armament consisted of a bow mounted MG 34 machine gun, a coaxial MG 34, and a MG 42 for use by the tank commander on an anti-aircraft mount attached to the cupola. A total of 5,850 rounds of 7.92MM ammunition was carried.

Tiger IIs converted to the command tank role had extra radios and antennas installed in the turret, reducing the ammunition stowage for the main gun. The command tanks were designated *Panzerbefehlswagen mit 8.8CM KwK43 L/71* or *Panzerbefehlswagen Tiger ausf B*.

The first Tiger IIs were issued to sPzAbt 503 during the late Spring of 1944, equipping the Headquarters section and the 1st Company. These early Tiger IIs of the 3rd Company are seen at gunnery training during August of 1944. The tanks are Dark Yellow with Olive Green and Red Brown sprayed patterns while the turret numbers are Black with White outlines.

The long barrel of the 88MM L/71 KwK43 gun of the Tiger II ausf B overhung the front of the hull by a considerable distance and drivers had to exercise care when maneuvering their tanks to avoid digging the muzzle into the ground or hitting obstructions while turning the vehicle.

The early Porsche turret's front curved plate formed a shot trap that was a serious disadvantage for these vehicles. These early Tiger IIs have a complete coating of zimmerit anti-mine cement. The side skirts on the Tiger II were not thin sheet metal, but rather 5MM thick sheet that served as anti-bazooka skirts similar to those on late PzKpfw IVs.

Tiger II ausf Bs of the 1st company of sPzAbt 503 in Normandy are parked under the trees and have foliage applied to them in order to hide them from Allied aircraft. The Tiger II in the foreground has all the side skirts fitted; however, the Tiger II in the background has no skirts fitted.

Turret

**Tiger II Ausf B
Porsche Turret**

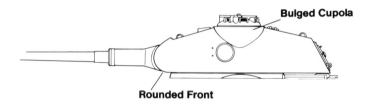

Bulged Cupola

Rounded Front

**Tiger II Ausf B
Henschel Production
Turret**

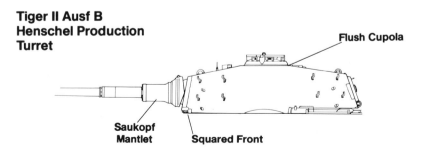

Flush Cupola

Saukopf
Mantlet

Squared Front

Tiger II 200 carries the standard three-color(Dark Yellow, Red Brown and Olive Green) camouflage scheme as did all the other Tiger IIs of sPzAbt 503. The vehicle carries Black turret numbers outlined in White. There was a good deal of variation in placement of the turret numbers within sPzAbt 503.

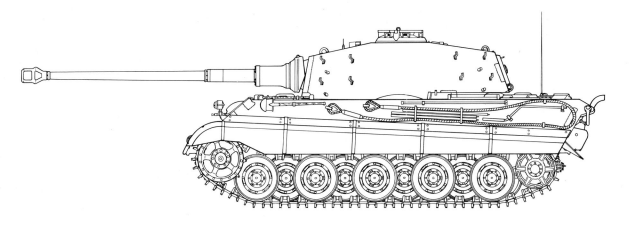

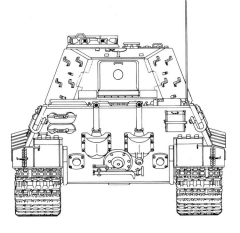

Specifications

Tiger II ausf B

Length . . . . 33 feet 8 inches
Width . . . . 12 feet 3 ⅝ inches
Height . . . . 10 feet 1 ⅝ inches
Weight . . . . 68 tons

Armament
   Main . . . . 88ᴍᴍ KwK 43 L.71 cannon
   Secondary . . . Three 7.92ᴍᴍ machine guns

Engine . . . . One 600 hp Maybach HL230 P30 gasoline engine.

Performance
   Speed . . . . . . 25.7 mph
   Range . . . . 106 miles (road) 75 miles (cross country)
Crew . . . . Five

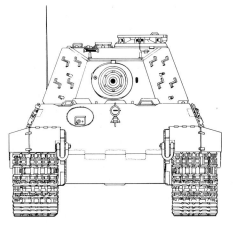

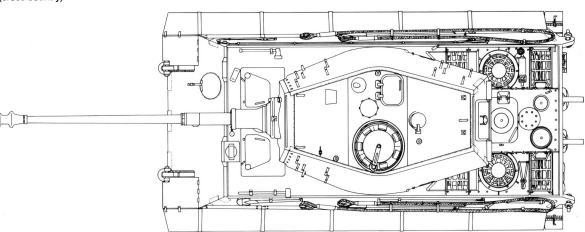

In September of 1944, sPzAbt 503 received new Tiger II ausf B tanks to replace losses and complete their conversion. This Tiger II has the Henschel production turret, which greatly improved turret armor protection over the Porsche design. Most vehicles were camouflaged by the receiving units, and the resulting patterns were usually dependent on the ability of the crew.

This Tiger II of SSsPzAbt 501 was being retrieved by U.S. Army engineers during the Battle of the Bulge, when it broke down and had to be abandoned. The turret numbers are Red and White, and the camouflage is Dark Yellow with Red Brown and darker than usual Olive Green. The unit symbol just visible to the left of the bow machine gun mount is in White. (U.S. Army)

Tank crews use a portable spray gun to paint the Olive Green and Red Brown camouflage finish to their Tiger II. The smaller straight turret front plate of the Henschel turret eliminated the shot trap of the Porsche turret, and the *Saukopf* (boar's head) mantlet was also an improvement over the mantlet used on the Porsche turret.

Most early Tiger IIs were destroyed in the retreat from Normandy during the Summer of 1944. This burned out Tiger II, with an early Porsche turret, was apparently being towed by the *Bergepanther* immediately in front of it. Despite the heavy foliage placed around both vehicles, they were located and destroyed.

Tiger II 233, was one of six Tiger IIs of sPzAbt 503 loaned to COL Otto Skorzeny for Operation *Panzerfaust*, the capture of the *Burgberg* in Budapest, Hungary. The engine deck air intake cover has a raised screen over it which improved engine cooling and the hull hatches on the Tiger II lifted and rotated to the side when opened.

The late war ambush dappled camouflage paint scheme was not used on Tiger IIs often. As these heavy tanks began being used in defensive positions, such patterns became more widely used. The dappled spots are Dark Yellow over the standard three-color camouflage scheme. This vehicle also still has its narrow transport tracks fitted. (Eric Helmuth)

The massive size of the Tiger II's main gun can be appreciated when compared to the crewmen next to it. The 88mm L/71 gun was probably the best weapon on any tank during the Second World War. The circular plate under the gun is the driver's compartment ventilator.

Climbing out to the end of the main gun's long barrel was the easiest way to install or remove the muzzle brake dust cover. The Bosch headlight, normally installed on the hull front plate, has been removed and stowed, a common practice during daylight hours.

Jagdtiger

German Army policy during mid-1943 was to develop a self-propelled gun mount based on each new tank chassis design. This allowed the Germans the capability of mounting a much larger weapon on the chassis than could be fitted into a fully rotating turret. When Henschel designed the Tiger II ausf B, the firm also cooperated with Krupp in designing the self-propelled gun version of the Tiger II.

A wooden full size model was displayed in October of 1943, and the first prototype appeared in April of 1944. Two examples were constructed with a Porsche designed suspension similar to that used on the *Ferdinand/Elefant* tank destroyers. While this suspension saved manufacturing time and internal hull space, its components were more highly stressed and, on at least one occasion, a complete bogie truck snapped off the hull during testing. To ensure that production vehicles would be rapidly available for service introduction, the standard Henschel torsion bar suspension was used for all production vehicles.

The new vehicle, designated at first *Jagdpanzer VI*, and then later the *Jagdtiger*, was assigned the type number SdKfz 186. It was basically a slightly lengthened Tiger II hull with a large box-shaped superstructure mounted in the space where the tank's turret had been previously mounted. The front armor was 250MM thick, while the sides and rear were 80MM thick. The hull had 100 to 150MM front armor, with 80MM sides and rear. The top and bottom plates were 40MM thick. Two large hinged doors were mounted in the rear superstructure armor plate and additional hatches, the ventilator, vision periscopes, and the gunner's sight were mounted in the roof of the fighting compartment. The driver and radio operator were both provided with standard Tiger II hull roof hatches.

The main armament for the *Jagdtiger* was the 128MM PaK44 L/55 gun, the largest weapon fitted to a production vehicle during the Second World War. Capable of destroying almost any Allied heavy tank at near-maximum range, this gun used separate ammunition to provide as much internal stowage as possible, with forty rounds of HE and AP ammunition being carried. Secondary armament consisted of an MG 34 machine gun in a standard Tiger II bow mount, and an MG 42 machine gun for anti-aircraft defense. The MG 42 was provided with an elevated gun mount which fitted into a mount on the rear engine deck.

The *Jagdtiger* was subject to frequent breakdowns, as the chassis and powertrain were severely overloaded by the vehicle's weight. Additionally, its size made it difficult to conceal from Allied aircraft, which had gained near total air superiority over Germany. The main gun was an excellent weapon, and those vehicles that were used from positions of good cover gave a good account of themselves. On the move, it was often a different story.

One American tank destroyer battalion in Belgium came across an abandoned *Jagdtiger* sitting across the road, and all attempts to move or tow the beast were completely unsuccessful. Even abandoned, this vehicle stopped the American advance until combat engineers could cut a road through the woods next to the abandoned German vehicle. Even though the *Jagdtiger* suffered from mechanical problems caused by its rushed development and a lack of a more powerful chassis, it was a real engineering achievement. The real failure of the German very heavy tanks/self-propelled guns was that they had exceeded the capabilities of German industry to produce them in the numbers needed and they exceeded the capacity of the German road and rail transport network to move them. Still, a large number of German tank design features found their way into many postwar Allied tank designs. Perhaps this is the best epitaph the Tigers could have.

A *Jagdtiger* of the 3rd company of JgTgrAbt 512, is examined by U.S. troops after it was blown up by its crew. The whole top turret roof plate is gone and the 128MM PaK44 main gun has been blown out of its mount. The colors are Dark Yellow with Red Brown over a base of Dark Gray. The side number, 332, is Black with a White outline. (U.S. Army)

U.S. troops examine an abandoned *Jagdtiger* before preparing it for shipment back to the rear during 1945. A very heavy layer of dust covers the vehicle obscuring the complex color scheme and Black and White vehicle numbers. (U.S. Army)

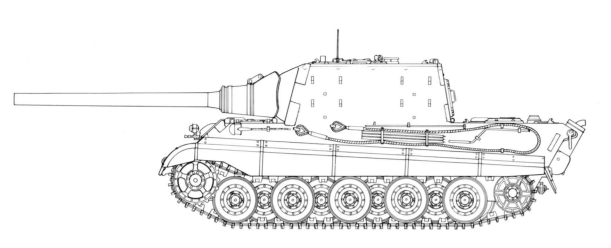

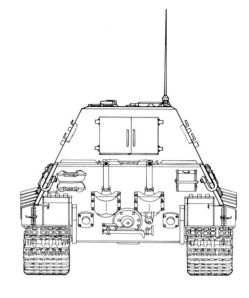

Specifications
Jagdtiger

Length . . . . 34 feet 11 ½ inches
Width . . . . 11 feet 10 ¾ inches
Height . . . . 9 feet 3 inches
Weight . . . . 70 tons

Armament
  Main . . . . 128MM PAK 44 L/55 cannon
  Secondary . . . Two 7.92MM machine guns.

Performance
  Engine . . . . . . One 600 hp Maybach HL230 P30
                  gasoline engine.
  Speed . . . . . . 23.6 mph
  Range . . . . 100 miles (road) 75 miles (cross country)
  Crew . . . . Six

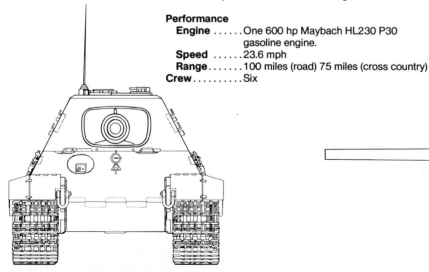

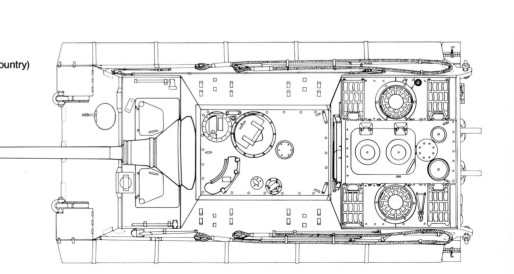

This pair of abandoned *Jagdtigers* were deserted by their crews during late 1945. The vehicle in the foreground is emplaced in a fairly good defensive position from which it could command the street. *Jagdtigers* were most often used in such positions to slow up the Allied advance. (U.S. Army)

One of the two experimental *Jagdtigers*, which were fitted with the Porsche suspension system, was issued to sPzJagAbt 653 and was later destroyed by its crew. The tremendous force of the demolition charge placed inside the vehicle has torn apart the superstructure, blown off the gun barrel, and thrown the engine completely away from the vehicle.

Suspension

**Jagdtiger
Early
Porsche
Suspension**

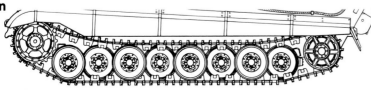

**Jagdtiger
Production
Suspension**

This *Jadgtiger* was captured by American troops in Germany during 1945 and returned to the United States for evaluation. This same vehicle can be seen today, displayed at the U.S. Army Aberdeen Proving Grounds. (U.S. Army)

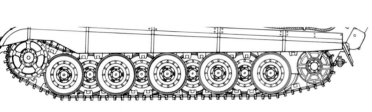

Panzer Colors

From Squadron/Signal

6251

6252

6253

squadron/signal publications